The Fundamental Ultimate Question: What Is?

A Philosophical Framework

By Corey Shepherd

ISBN (Paperback): 979-8-89122-577-0

Table of Contents

About the Author

Corey Shepherd is a philosopher, scientist, educator, and entrepreneur whose diverse life experiences have shaped his unique perspective on the nature of existence. A former U.S. Navy veteran who served ten years in tactical operations and as a military instructor, after initially struggling in traditional academic settings, Shepherd discovered the transformative power of adaptability and lifelong learning.

Following his military service, Shepherd earned his Bachelor's degree in business management and spent eight years in corporate leadership before transitioning to entrepreneurship. He currently teaches multimedia and business concepts to high school students while running his own business, combining his practical experience with his passion for education and mentorship.

Beyond his professional pursuits, Shepherd is also a singer-songwriter and author, having written two books that explore the interconnected nature of existence. His philosophical insights emerged from navigating significant personal challenges, leading him to recognize the profound and unexpected ways in which seemingly disparate aspects of reality connect to form a unified whole.

Shepherd's First-Principles approach to understanding reality is deeply interrelational, drawing connections between his military training, business leadership experience, creative pursuits, and spiritual inquiry. This multifaceted background allows him to bridge practical wisdom with philosophical depth, making complex concepts accessible to diverse audiences.

"The Fundamental Ultimate Question: What Is?" represents the culmination of Shepherd's journey from academic struggles to profound understanding, offering readers the same framework for peace and clarity that transformed his own life through periods of difficulty and growth.

Page Left Blank Intentionally

Along the journey to understand reality more deeply, we all experience great pain and many pleasures. Mistakes are made, and lessons are learned. I have struggled in my own ways, such as a lack of confidence and acceptance of other perspectives. My struggles dissolved when I reached the following conclusions about the deepest concepts of reality. Since that moment, I have moved with an improved sense of grace and profound peace, which I apply toward the successful navigation of this human condition. My hope is that this worldview inspires every reader to discover lasting peace in their own way.

Introduction

The question "What is?" appears deceptively simple, yet it may contain within it the entire mystery of existence. This work will guide you through the layers of perception that shape your reality, revealing the profound limitations that define human understanding while pointing toward what may be the infinite nature of that which lies beyond.

We exist as conscious observers in a universe that appears to be attempting to understand itself through every possible perspective. You may not be separate from this cosmic self-inquiry; you might *be* it. Through exploring the boundaries of perception, we discover that reality seems to operate as a complex geometric hologram, where consciousness manifests as nodal points of awareness within what could be the fundamental substrate of existence.

Note on approach: While this exploration draws upon established scientific principles, the ultimate conclusions should be understood as plausible inferences that emerge from connecting diverse observations. The logical gaps between scientific facts and metaphysical implications are acknowledged throughout, and readers are invited to follow the reasoning while maintaining healthy skepticism about the final claims.

Chapter 1: Ranges of Perception

Imagine standing at the edge of an ocean. The horizon appears as a definite line where water meets sky, yet you know this boundary exists only from your particular vantage point. Step closer to the water, and the horizon shifts. Climb a cliff, and it expands. This simple observation reveals a profound truth: what we perceive as the limits of reality are more accurately defined as the limits of our perception.

Your sensory apparatus creates a bubble of awareness around you. Within this bubble, you navigate daily life with confidence, making decisions based on the information available to your particular range of perception. But consider this: the electromagnetic spectrum extends far beyond visible light, sound waves carry information you cannot hear, and quantum interactions occur beneath the threshold of your direct observation.

The visible light that creates your visual experience represents only about 0.0035 percent of the total electromagnetic spectrum that surrounds you constantly. Beyond the red and violet boundaries of your vision lie vast realms of infrared, ultraviolet, microwaves, radio waves, X-rays, and gamma rays, entire universes of electromagnetic phenomena that remain invisible to your naked perception.

Other creatures inhabit different portions of this spectrum. Bees see ultraviolet patterns on flowers that guide them to nectar. Snakes detect infrared radiation that allows them to hunt warm-blooded prey in total darkness. Sharks sense the electrical fields generated by living creatures. Each species has evolved to perceive the electromagnetic information most relevant to its survival needs.

The boundary of your perceptual bubble is not fixed. With instruments, you can extend your sight into the microscopic realm or peer into distant galaxies. Yet even these extensions have limits. There remains what appears to be an infinite expanse of reality that lies forever beyond your ability to perceive directly.

Why does this matter? Because recognizing the boundaries of your perception is the first step toward intellectual humility. It opens the possibility that reality may contain dimensions, forces, and forms of existence that your current understanding cannot encompass. This recognition creates space for wonder and prevents the premature closure of inquiry that comes from assuming your perceptual bubble contains all that is.

The mystics and philosophers throughout history have pointed toward this truth: the known is finite, the unknown appears to be infinite. By acknowledging the vast unknown that surrounds your small sphere of direct experience, you begin to participate in the larger mystery of existence rather than remaining trapped within the narrow confines of assumed certainty.

Chapter 2: Scalability

Stand before a mirror and observe your reflection. You see a human being of particular height and weight, existing at a scale that feels natural and central to your experience. Yet this feeling of centrality may be an illusion born of your perceptual limitations.

Consider the journey inward: your body contains trillions of cells, each cell contains organelles, each organelle contains molecular machinery, each molecule contains atoms, each atom contains subatomic particles, and the exploration continues down to quarks, leptons, and the fundamental quantum fields that underlie all matter. At each level, entire universes of activity unfold, completely invisible to your naked eye.

Now consider the journey outward: you exist on a planet orbiting a star within a galaxy containing hundreds of billions of stars, and this galaxy is one among trillions in an observable universe that may itself be only a small part of a larger multiverse. The scales involved stretch beyond human comprehension.

Plot yourself on this scale from the smallest known structures to the largest, and a remarkable pattern emerges: you exist roughly in the middle. This is not a coincidence but a consequence. Your perceptual apparatus evolved to function effectively at the scale where human survival depends on successful navigation. You are naturally attuned to objects and processes that exist at scales similar to your own body.

This central positioning reveals something profound about the nature of perception itself. You are not viewing reality from some objective, universal standpoint. You are embedded within reality at a particular scale, and your understanding is necessarily limited by this embedding. The very instruments you use to extend your perception are built from materials and processes that exist at scales accessible to your manipulation.

In Jason Padgett's Quantum Information Holography (QIH) theory, this scaling takes on deeper meaning. Reality may be structured as information encoded on what he calls "Planck qubits," fundamental building blocks made of light that form a lattice at the smallest possible scale. As light traverses this lattice, it encodes data through the "angles of light relative to the spacetime grid," defining velocity, acceleration, gravity, and even quantum probabilities.

Understanding your position in the scale of existence cultivates appropriate humility. The vast majority of reality appears to operate at scales either too small or too large for your direct perception. Recognizing this limitation opens you to the possibility that the universe may contain forms of organization, intelligence, and beauty that exist beyond your current ability to perceive or comprehend.

Chapter 3: Categorization

Your mind is a categorizing machine. From the moment you wake until you sleep, you sort experiences into categories: friend or stranger, safety or danger, pleasant or unpleasant, important or irrelevant. This categorization process allows you to navigate complexity by reducing the infinite detail of experience to manageable units of understanding.

Consider how you categorize people: by age, appearance, profession, personality traits, relationship to you, and countless other dimensions. You might categorize potential romantic partners by compatibility factors, friends by shared interests, or colleagues by professional competence. Each categorization system serves a purpose, helping you make decisions and predictions about behavior.

The same process extends to every aspect of experience. You categorize foods by taste, texture, and nutritional value. You categorize activities by difficulty, enjoyment, and time requirements. You categorize information by reliability, relevance, and actionability. Your entire conceptual framework consists of interlocking categorization systems.

But here's the crucial insight: your categories are not objective features of reality. They are constructions of your particular nervous system, shaped by your evolutionary history, cultural conditioning, and personal experience. A different type of consciousness might categorize the same phenomena in completely different ways.

From an information-theoretic perspective, categorization represents a compression algorithm that reduces the infinite complexity of reality to patterns your finite consciousness can process. As John Wheeler suggested with his "it from bit" hypothesis, if reality is fundamentally informational, then your categories are ways of parsing the vast data stream of existence into meaningful chunks.

Recognizing the constructed nature of your categories serves several purposes. First, it increases your flexibility. When you understand that your categories are tools rather than truths, you can modify them when they no longer serve your purposes. Second, it reduces judgment. Understanding that others operate with different categorization systems makes their behavior more comprehensible, even when it differs from your own.

Most importantly, recognizing the limitations of categorization opens you to the possibility that reality itself may transcend all categories. The fundamental nature of existence might be too fluid, too complex, or too unified to be captured by any system of classification. This recognition points toward a more direct, immediate relationship with what is, unconditioned by the filtering effects of conceptual frameworks.

Chapter 4: Complexity

Complexity surrounds you, but your perception naturally simplifies. When you look at a forest, you see trees, not the intricate ecosystem of soil microorganisms, fungal networks, insect communities, and biochemical processes that actually sustain the forest. When you observe a conversation, you notice words and facial expressions, not the neurochemical cascades, social dynamics, and cultural patterns that shape every interaction.

This simplification is necessary for survival. If you attempted to consciously process every detail of every system you encounter, you would be overwhelmed into paralysis. Your nervous system evolved to filter complexity, extracting only the information necessary for effective action at your particular scale of existence.

Yet this filtering process has profound implications. The vast majority of what is actually happening in any moment remains invisible to your direct perception. The complexity of even the simplest phenomena, a single cell dividing, a photon interacting with matter, a neural network processing information, exceeds your ability to comprehend fully.

Consider global systems: climate patterns, economic networks, technological infrastructures, and social institutions. Each represents a level of complexity that no single mind can grasp completely. You can understand pieces, patterns, and general principles, but the full dynamic reality of these systems remains beyond individual comprehension.

This limitation extends to your own inner experience. The complexity of your own nervous system, with its billions of neurons forming trillions of connections, creates patterns of activity that generate your thoughts, emotions, and perceptions. Yet you have no direct access to this underlying complexity. You experience its products, thoughts, and feelings, but not the intricate processes that create them.

In QIH theory, this complexity emerges from simple rules governing light and information at the quantum level. Padgett describes reality as emerging from “a profound interplay between light, information, and probability that echoes the essence of our universe.” What appears as irreducible emergent complexity at your scale may result from elegant simplicity at the fundamental level.

Recognizing the limits of your ability to perceive complexity cultivates appropriate humility. It suggests that reality may operate according to principles and patterns that exceed your current understanding. This recognition opens space for wonder and prevents the premature closure that comes from assuming your simplified perceptions capture the full nature of what is.

Chapter 5: Dimensionality

You exist in three spatial dimensions plus time, but this familiar framework may represent only a fragment of reality's true dimensionality. Consider how a two-dimensional being living on a flat surface would experience your three-dimensional world. As you moved through their plane, you would appear as a changing cross-section, a mysterious shape that morphs and transforms according to no logic apparent to the two-dimensional observer.

Similarly, higher-dimensional phenomena may be constantly intersecting your three-dimensional space, appearing as seemingly disconnected events or patterns that make no sense from your limited dimensional perspective. Modern physics suggests that reality may contain ten, eleven, or even infinite dimensions, most of which remain completely inaccessible to your direct perception.

This dimensional limitation affects not just your perception of space but also your understanding of causality, relationships, and possibility. In higher dimensions, points that appear separate in your three-dimensional space might be intimately connected. Actions that seem to have no relationship from your perspective might be aspects of a single higher-dimensional process.

The mathematical frameworks that describe these higher dimensions remain largely abstract to human consciousness. You can manipulate the equations, but you cannot visualize or directly experience what they represent. This represents a fundamental limitation: your consciousness appears to be structured in a way that can only navigate a small subset of reality's total dimensionality.

Yet this limitation points toward something profound. If consciousness itself is embedded in a higher-dimensional reality, then your individual perspective represents one particular trajectory through a vast multidimensional space. Your life, your thoughts, and your experiences, all might be aspects of a higher-dimensional pattern that connects you to the totality of existence in ways that transcend your current ability to perceive or understand.

In the holographic principle from physics, all information contained in a volume can be encoded on its boundary. This suggests that what you experience as three-dimensional reality might be a projection from information encoded in higher dimensions. Your consciousness might be reading a holographic projection of a much more complex higher-dimensional reality.

Chapter 6: Geometry and Information

Reality appears to be structured geometrically, but not in the simple way you might expect. The geometry of existence operates at levels both more fundamental and more complex than the shapes you learned in school. At the quantum level, particles exist in superposition states that defy classical geometric description. At the cosmic level, space itself curves and warps in ways that create the phenomenon you experience as gravity.

Between these extremes, geometric principles organize phenomena in ways that suggest deep underlying order: the spiral patterns in shells, galaxies, and DNA; the hexagonal structures in crystals, honeycombs, and molecular arrangements; the fractal patterns that repeat at different scales throughout nature. These geometric regularities hint at fundamental organizing principles that shape reality at every level.

Cymatics, the study of sound's effects on matter, reveals how vibration creates geometric patterns. When sound waves interact with granular materials, they spontaneously organize into complex geometric forms. This suggests that the geometric structures you observe in nature may be the visible manifestation of underlying vibrational patterns.

Sacred geometry traditions across cultures have recognized these patterns as more than mere mathematical curiosities. They may represent the fundamental architecture of existence itself. The Golden Ratio, the Flower of Life, and the Platonic Solids are geometric forms that appear repeatedly in natural phenomena. They have been recognized as keys to understanding the structural principles of reality.

In QIH theory, this takes on computational significance. As Padgett describes, light encodes information through its angles relative to the spacetime grid of Planck qubits. "The angle of light, theta (θ), relative to the grid becomes a crucial parameter in defining velocity," and other fundamental properties. Geometry becomes the language through which information is processed at the most fundamental level.

Your own consciousness may be geometrically structured. The way you organize experience, the patterns of your thoughts, the architecture of your memories, all might reflect geometric principles that operate at the level of consciousness itself. This suggests that understanding geometry is not merely an intellectual exercise, but a way of aligning your awareness with the fundamental structure of existence.

Chapter 7: Electromagnetism and Spectral Reality

You see only a tiny sliver of reality's electromagnetic spectrum. The visible light that creates your visual experience represents less than one percent of the total electromagnetic radiation that surrounds you constantly. Beyond the red and violet boundaries of your vision lie vast realms of infrared, ultraviolet, microwaves, radio waves, X-rays, and gamma rays—entire universes of electromagnetic phenomena that remain invisible to your naked perception.

Your technological extensions allow you to translate other portions of the electromagnetic spectrum into forms your senses can process. Radio telescopes capture cosmic radiation and convert it into images. Infrared cameras reveal heat signatures. X-ray machines penetrate solid matter. Yet even these extensions represent only a fraction of the electromagnetic phenomena that exist.

The electromagnetic spectrum provides a perfect metaphor for the nature of reality itself. What you directly perceive represents only a narrow band of what actually exists. The vast majority of electromagnetic reality remains invisible to you, yet it continues to influence your world in countless ways. Radio waves carry information through the air around you. Cosmic radiation constantly bombards your planet. Magnetic fields generated by your planet's core protect you from solar radiation.

This invisible electromagnetic environment shapes your reality in ways you rarely consider. Your brain operates through electrical impulses. Your heart generates electromagnetic fields. Your nervous system responds to electromagnetic influences. You are not separate from the electromagnetic universe; you are a complex electromagnetic phenomenon embedded within a much larger electromagnetic reality.

In QIH theory, this electromagnetic foundation becomes the basis for reality itself. Information is encoded in light, electromagnetic radiation, as it traverses the quantum lattice. What you experience as matter, energy, space, and time may all emerge from the processing of electromagnetic information at the Planck scale.

Chapter 8: Spectral Fractalization and Emergence

Reality exhibits fractal properties, patterns that repeat at different scales and levels of complexity. Zoom into a coastline, and you discover that its irregular boundary contains smaller irregularities, which contain still smaller ones, continuing indefinitely. This fractal structure appears throughout nature: in the branching of trees, the formation of clouds, the distribution of galaxies, and the organization of neural networks.

Fractalization suggests that reality is not organized in simple hierarchical levels but rather in complex, self-similar patterns that emerge at every scale. This is a concept we might call "spectral fractalization." the emergence of similar patterns across different spectra of existence: spatial, temporal, energetic, and informational.

Emergence occurs when complex systems develop properties that cannot be predicted from understanding their individual components. A single neuron cannot think, but networks of neurons give rise to consciousness. Individual water molecules do not exhibit wetness, but collections of them create the phenomenon you experience as liquid. Simple rules governing individual behavior can generate complex collective patterns.

The relationship between fractalization and emergence creates a reality that appears to be both unified and infinitely complex. The same fundamental patterns repeat at every scale, yet at each level of organization, new properties emerge that transcend the sum of their parts. This suggests that reality may operate according to principles that are both elegantly simple and incomprehensibly complex.

Your consciousness participates in this fractal, emergent reality. The patterns of your thoughts may reflect the same organizational principles that structure galaxies. The emergence of your subjective experience from neural activity represents the same type of phase transition that occurs when matter organizes into life, or when life organizes into consciousness.

Understanding fractalization and emergence provides a framework for comprehending your place in the larger patterns of existence. You are not a separate entity observing reality from the outside. You may be an emergent property of reality's fractal organization, a node of consciousness through which the universe comes to know itself.

Chapter 9: Time and Eternalism

Time is not what it appears to be. Your experience of time as a linear progression from past through present to future may represent a perceptual construct rather than an objective feature of reality. Einstein's revelations about relativity demonstrate that time is relative to the observer's motion and gravitational field. Time passes differently for different observers, and under extreme conditions, time can slow down or even theoretically approach infinity.

At the speed of light, time dilation becomes infinite, and time appears to stop. This suggests that, from the perspective of a photon, all of existence may occur in a single, eternal now. Past, present, and future could collapse into simultaneity. The linear sequence of events that structures your experience might dissolve into a unified, timeless reality.

Physicist Julian Barbour has developed this insight into a comprehensive theory. Barbour argues that "we live in a universe which has neither past nor future" and that "our sense of the passage of time is nothing more than a giant cosmic illusion." In his view, "time as such does not exist. There is no invisible river of time. But there are things that you could call instants of time, or 'Nows.'" Much like the Planck-Qubits found in QIH, the smallest increments of time may be considered as Time-Bits.

According to Barbour, these "Nows" are "arrangements of everything in the universe relative to each other in any moment." He calls his timeless universe "Platonia" after Plato's world of eternal forms, where "all the arrangements of matter that can be" exist simultaneously in a vast, multidimensional space. This concept is similar to shining a flashlight on the ground at night. You can only see what is illuminated by the light, yet the remaining landscape exists beyond immediate perception.

This has profound implications for understanding the nature of existence. If time is relative and can potentially approach infinity, then the linear progression that defines your experience of causality, change, and development may be only one way of organizing a more fundamental timeless reality. Everything that you perceive as happening sequentially may actually exist simultaneously in a higher-dimensional space that transcends temporal sequence.

Your consciousness appears to be structured in a way that creates the experience of temporal flow. You remember the past, anticipate the future, and experience the present moment as a moving boundary between what was and what will be. Yet this temporal structure may be a feature of your particular type of consciousness rather than an objective property of reality itself.

If all moments exist simultaneously in a timeless reality, then your linear experience represents a particular trajectory through this eternal landscape. Your life, your thoughts, and your experiences, all exist as permanent features of reality's timeless structure. Nothing is truly lost, because everything exists eternally in the fundamental substrate of existence.

Chapter 10: Sensory Input and Perception

Your five senses create a limited interface with reality. Vision, hearing, touch, taste, and smell provide you with information about your environment, but each sense operates within narrow parameters. Your eyes detect only a small portion of the electromagnetic spectrum. Your ears hear only a limited range of sound frequencies. Your other senses operate within similarly restricted ranges.

Consider the complexity of a single sensory experience. When you hear a musical note, you are actually perceiving a complex wave pattern that contains not just the fundamental frequency, but also harmonics, overtones, and subtle modulations that give the note its particular character. Your auditory system extracts meaningful information from this complex pattern, allowing you to recognize melodies, distinguish instruments, and respond emotionally to music.

This same principle applies to all sensory experiences. What appears to be simple sensory input actually involves complex processing that extracts meaning from vast amounts of information. Your nervous system filters, interprets, and constructs your perceptual experience from the raw data provided by your sensory organs.

Beyond the traditional five senses, you possess additional sensory capabilities that often go unrecognized. Your sense of balance, your awareness of your body's position in space, your ability to sense time passage, and your capacity to detect subtle changes in your internal state all provide information about your relationship to your environment.

Yet even these extended sensory capabilities represent only a fraction of the information available in your environment. Many creatures possess sensory abilities that exceed your own: echolocation, magnetic field detection, electrical field sensing, and many others. The sensory world inhabited by other species remains largely inaccessible to your direct experience.

Thought itself might be considered a sixth sense, a way of receiving information about the patterns and relationships that exist within your experience. Your thoughts provide you with information about meaning, significance, and connection that transcends what your physical senses can detect directly. And, much like vision or hearing, you cannot control that which enters your perception through thought. Nor can any individual process the entire spectrum of thought. You can only react to it once the information is processed.

In the context of QIH theory, your senses may be tuned to specific frequencies of the electromagnetic information that creates reality. What you perceive as solid matter, sound, heat, and light may all be different ways of experiencing the electromagnetic information processing that underlies existence.

Chapter 11: Emotions as Informational Tools

Emotions represent another dimension of your perceptual apparatus. They provide information about your relationship to your environment, your internal state, and the significance of events in your life. Fear alerts you to danger. Joy indicates alignment with your deeper values. Sadness signals loss or disconnection. Anger mobilizes energy for action or boundary setting.

The range of possible emotional experiences extends far beyond what any individual consciousness can encompass. Different cultures, different species, and different individuals within the same species experience emotional ranges that vary dramatically. Your particular emotional capacity represents one small subset of the total emotional spectrum that may exist in the universe.

Emotions emerge from the complex interaction between your nervous system, your environment, and your interpretive frameworks. They are not simply reactions to external events, but complex constructions that involve memory, anticipation, cultural conditioning, and personal history. The same event can evoke completely different emotional responses in different individuals or even in the same individual at different times.

Your emotional experience provides valuable information about your alignment with your environment and your deeper nature. Emotions that feel harmonious and expansive may suggest alignment with your natural patterns. Emotions that feel discordant or constrictive might indicate areas where adjustment is needed.

Understanding emotions as information rather than as problems to be solved opens new possibilities for a relationship with your internal experience. Rather than judging emotions as good or bad, you can learn to read them as sophisticated feedback about your relationship to reality. This creates the possibility of emotional wisdom, the ability to receive and respond to emotional information skillfully.

From an information-theoretic perspective, emotions may represent sophisticated data processing about the patterns and relationships in your environment. They could be rapid, intuitive computations about complex social and environmental variables that your conscious mind cannot process directly.

Chapter 12: Mindset and Reality Construction

Your mindset shapes your reality more than you might realize. The interpretive framework you bring to experience determines what you notice, how you understand what you observe, and what possibilities you can imagine. Different mindsets literally create different realities from the same underlying phenomena.

Consider the difference between a victim mindset and a response-ability mindset. From a victim perspective, challenges represent evidence of an unfair world where external forces control your fate. From a response-ability perspective, the same challenges represent opportunities to develop resilience, creativity, and wisdom. The external situation remains the same, but the lived experience differs dramatically.

Emergency situations reveal the life-or-death importance of mindset. In a crisis, those with survival mindsets focus on immediate actionable steps, while those with victim mindsets may become paralyzed by the overwhelming nature of the situation. The same crisis can lead to death or survival depending on the mindset of the individual involved.

Your mindset is not fixed. It represents a learned pattern of perception and interpretation that can be consciously modified. By observing the effects of different mindsets, you can choose to cultivate those that serve your well-being and discontinue those that create suffering.

The range of possible mindsets extends far beyond what most people explore: growth mindset versus fixed mindset, abundance mindset versus scarcity mindset, curiosity mindset versus judgment mindset. Each represents a different way of organizing your relationship to experience.

Understanding the power of mindset reveals your participation in creating your reality. You are not a passive observer of a fixed external world. You are an active participant whose mindset shapes both what you perceive and how you respond to what you perceive. This recognition empowers you to take responsibility for your experience and to consciously choose mindsets that align with your deepest values and aspirations.

If reality is fundamentally informational, as suggested by QIH and information theory, then your mindset functions as a filter and interpreter of the information stream that constitutes your experience. Different mindsets literally tune you to different aspects of the infinite information field that surrounds you, and through training, experience, and intention, your mindset is within your control.

Chapter 13: Morality and Vibrational Harmony

Morality is not absolute but relational. What benefits one aspect of existence may harm another. The spider's survival depends on the fly's death. The plant's growth requires the decomposition of organic matter. The immune system's health depends on the destruction of pathogens. Good and evil are not cosmic principles but descriptions of relationships between different aspects of existence.

This relational understanding of morality extends to human affairs. Actions that benefit one group may harm another. Decisions that serve short-term interests may damage long-term well-being. Behaviors that support individual development may conflict with collective needs. Moral complexity arises from the intricate web of relationships that connect all aspects of existence.

From a vibrational perspective, good and evil might be better understood as harmony and discord. Harmonious vibrations amplify and support each other, creating constructive interference patterns. Discordant vibrations cancel and conflict, creating destructive interference patterns. What you experience as pleasure and pain may be your consciousness detecting these harmonic relationships.

Your moral intuitions evolved to support the survival and thriving of your particular form of consciousness within your specific environment. These intuitions provide valuable guidance, but they are not universal truths. They represent one perspective within the vast web of perspectives that constitute reality.

Understanding morality as relational rather than absolute creates space for both compassion and wisdom. Compassion arises from recognizing that all beings may be attempting to survive and thrive within their particular circumstances. Wisdom arises from recognizing the complex web of relationships that connect all actions to their consequences.

This understanding does not lead to moral relativism but to moral sophistication. By recognizing the relational nature of morality, you can make more skillful choices that account for the complex web of relationships that your actions affect.

Chapter 14: Expanding Consciousness

Self-awareness exists in degrees and dimensions that most people never fully explore. Maslow's hierarchy of needs provides one framework for understanding the levels of self-awareness. At the survival level, self-awareness focuses on basic needs and threats. At the self-actualization level, self-awareness encompasses purpose, meaning, and contribution to the larger whole.

Beyond individual self-awareness lies collective self-awareness, the recognition that your individual consciousness exists within larger patterns of consciousness. Family consciousness, cultural consciousness, species consciousness, and, potentially, cosmic consciousness represent expanding circles of self-awareness that transcend individual boundaries.

Your self-awareness is not fixed but dynamic. It can expand or contract depending on your circumstances, your practices, and your intentional development. Practices like meditation, self-reflection, and conscious relationships can expand your self-awareness. Trauma, addiction, and unconscious patterns can contract it.

The boundaries of your self-awareness define the boundaries of your ability to make conscious choices. Within the sphere of your self-awareness, you can respond consciously to circumstances. Beyond these boundaries, you react automatically according to unconscious patterns.

Expanding self-awareness is not merely personal development but may be a participation in the universe's attempt to understand itself. Your individual consciousness represents one node in the vast network of consciousness that may pervade reality. By developing your self-awareness, you might contribute to the collective awakening of consciousness itself.

Yet self-awareness also has limitations. Complete self-awareness may be impossible for individual consciousness. The complexity of your own neural processes, the influences of your unconscious patterns, and the web of relationships that shape your experience may exceed your capacity for conscious understanding.

In the QIH framework, expanding self-awareness might correspond to tuning into broader frequencies of the information field that constitutes reality. Greater consciousness could mean accessing more of the electromagnetic spectrum of information that surrounds you.

Chapter 15: Limitations of Perspective

Perspective is everything, yet no single perspective can encompass everything. Your viewpoint determines what you can see, what remains hidden, and how you interpret what you observe. Every perspective reveals certain aspects of reality while simultaneously concealing others.

The challenge of perspective lies in its invisibility to itself. You cannot see your own perspective directly; you can only see through it. This creates a fundamental limitation: you cannot know what you are missing from your current viewpoint. The blind spots created by your perspective remain invisible to you.

Understanding perspective requires recognizing both its power and its limitations. Your perspective shapes your reality in fundamental ways, determining what appears possible, what seems important, and what feels meaningful. Yet your perspective is also provisional and limited, representing one viewpoint among infinitely possible viewpoints.

The further you move from your own perspective, the more difficult it becomes to understand and empathize with other viewpoints. Cultural perspectives, species perspectives, and radically different forms of consciousness may be so alien to your own that you cannot even imagine what reality looks like from those standpoints.

Yet the recognition of perspective's limitations opens possibilities for expansion. By consciously exploring different perspectives, you can expand your understanding of reality's complexity. By recognizing the validity of other viewpoints, you can develop greater humility and compassion.

The ultimate perspective might be the recognition that all perspectives are partial and provisional. Reality itself may transcend any single viewpoint, containing infinite possibilities for perception and understanding. This meta-perspective creates space for both confidence in your own experience and openness to the vast unknown that lies beyond your current understanding.

In the context of higher-dimensional reality, your perspective might be analogous to a two-dimensional being trying to understand a three-dimensional object by observing its cross-sections. What seems like disconnected, mysterious phenomena may actually be fragments of a unified higher-dimensional reality beyond your direct perception.

Chapter 16: Classical vs. Quantum Physics

Classical physics describes the reality you can perceive directly. Objects have definite positions, velocities, and properties. Causes produce effects in predictable ways. Time flows linearly from past to future. Space provides a fixed container for physical events. These principles work perfectly for navigating the world at human scales.

Quantum physics describes the reality that may underlie your perceptual world. At the quantum level, particles exist in superposition states, occupying multiple positions simultaneously until observation collapses them into definite states. Quantum entanglement creates instantaneous connections between particles regardless of distance. The act of observation affects the reality being observed. Uncertainty becomes a fundamental feature rather than a limitation of knowledge.

The boundary between classical and quantum physics corresponds roughly to the boundary of your direct perception. Classical physics describes the world as it appears to your senses. Quantum physics describes the world as it may actually operate beneath the threshold of your perceptual apparatus.

This transition from classical to quantum reveals something profound about the nature of reality. The solid, predictable world of classical physics emerges from the fluid, probabilistic world of quantum mechanics. The definite properties you observe in classical objects arise from the indefinite potentials of quantum systems.

Your consciousness may operate according to quantum principles even though your experience appears classical. The hard problem of consciousness, how subjective experience arises from objective neural processes, might be solved by recognizing that consciousness itself operates quantum mechanically while creating the emergent illusion of classical experience.

Understanding the relationship between classical and quantum physics provides a framework for understanding your place in reality. You exist at the boundary between the quantum realm of infinite potential and the classical realm of definite experience. Your consciousness may be the process through which quantum potentials collapse into classical realities.

In QIH theory, this relationship becomes more explicit. The classical reality you experience emerges from quantum information processing at the Planck scale. In this view, your consciousness might be reading the classical output of quantum computations occurring in the fundamental structure of spacetime.

Chapter 17: The F.U.Q. - A Convergent Answer

Drawing together all these threads, the ranges of perception, scalability, complexity, time, quantum mechanics, consciousness, and information theory, a picture emerges that may answer the Fundamental Ultimate Question.

The answer that appears to emerge is that **it all may be now**. Everything that has ever existed, everything that will ever exist, and everything that could ever exist, all of it appears to be, not was or will be, but **is**. The linear time that structures your experience may be a perceptual construct that allows finite consciousness to navigate the eternal "nowness" of existence.

How this may work: At the most fundamental level, time appears to dissolve into timelessness. Past, present, and future may collapse into simultaneity. As Barbour suggests, there are "only timeless 'nows,'" existing as "arrangements of everything in the universe relative to each other." All possibilities may exist simultaneously in the infinite potential of quantum reality. Your individual consciousness represents one trajectory through this eternal landscape, creating the illusion of temporal sequence by following a particular path through the timeless whole.

Supporting evidence suggests:

- Einstein's relativity shows time is observer-dependent, with time dilation approaching infinity at light speed.
- Quantum mechanics reveals an underlying probabilistic reality where all possibilities may exist in superposition.
- Information theory suggests reality may be computational, with consciousness arising from information integration.

- The QIH proposes that reality emerges from light-encoded information at the Planck scale.
- Barbour's timeless physics provides mathematical frameworks for understanding reality without fundamental time.

Practical implications: Because everything may be now, nothing is ever truly lost. Your loved ones, your experiences, your contributions to the world, all may exist permanently in the eternal structure of reality. Death might not be the end of existence, but a transition from one form of consciousness to another. Your essential being remains as a permanent feature of what is.

Because you may be a nodal point of perception in the complex geometry of universal awareness, you might have the power to navigate your experience consciously. You could align your consciousness with harmonious patterns and manifest positive outcomes through conscious choice and action.

Most importantly, understanding what is may create the foundation for profound acceptance. You need not judge or resist what appears in your experience. Everything that arises may be an aspect of the universe's attempt to understand itself. Good and bad, pleasure and pain, success and failure, all may be equally valid expressions of the totality of existence.

This acceptance is not passive resignation but active participation in the cosmic process of self-discovery. By accepting what is, you may align yourself with the fundamental nature of reality. By navigating consciously within this acceptance, you might participate in the universe's evolution toward greater awareness and understanding.

Chapter 18: What Is?

Of the inquisitive terms (who, what, when, where, why, and how), "what" appears to be the most generalized, thereby offering the highest level of universal applicability. Stemming from "what," you can achieve the goal of all remaining inquisition.

- What is the reason for... (why)
- What is the location of... (where)
- What is the name of... (who)
- What is the point in time... (when)
- What is the process of... (how)

Yet "what," by itself, is incomplete. You cannot explore the nature of existence until you add the word "is." By asking "what is," you begin to provide context and a point from which every question can stem.

So, what is? Everything. Nothing. The question and the answer. The seeker and the sought. The observer and the observed. The boundary between "is" and "is not" may dissolve in the recognition that all apparent separations are constructions of limited perception.

You may not be separate from what you seek to understand. Consider the linguistic label humanity has assigned to the container of everything, the universe. Breaking that word down to its parts reveals the answer in and of itself. "Uni-" meaning one and "-verse" meaning song. You might be the universe, or one song, understanding its frequency interactions through the particular lens of your individual consciousness. Every question you ask, every answer you discover, every moment of confusion, and every flash of insight, all may be expressions of reality's fundamental nature exploring itself.

The journey through perception's limitations reveals not the inadequacy of your understanding but the possibly infinite nature of what is to be understood. Each boundary you discover points beyond itself to vast realms of possibility. Each limitation you recognize opens new dimensions of exploration.

This understanding may provide a foundation for living with both confidence and humility. Confidence because you might recognize your fundamental unity with the source of all existence. Humility because you recognize the infinite mystery that surrounds and permeates your individual perspective.

The enlightenment that may come from understanding what is does not eliminate the challenges of human existence. Pain still hurts. Loss still grieves. Confusion still confuses. But these experiences might be held within a larger context of meaning and purpose. They could become expressions of the universe's total experience rather than personal problems to be solved.

This context may transform suffering from meaningless affliction into meaningful participation in the cosmic process. Your struggles become the universe's struggles. Your insights become the universe's insights. Your growth becomes the universe's growth.

Living in the eternal now: If this understanding is correct, how do we live with such knowledge? The practical wisdom that emerges is elegantly simple: accept what is, navigate consciously within that acceptance, and recognize your fundamental unity with the source of all existence. This may be the key to emotional stability, effective action, and profound peace.

Addressing common concerns:

Does this eliminate free will? Not necessarily. Your choices may be real and meaningful precisely because they are part of the eternal structure of what is. Free will exists within the timeless landscape; you navigate consciously through infinite possibilities.

Does this make life meaningless? Quite the opposite. If you are the universe experiencing itself, then every moment of your life contributes to the cosmic understanding. Your individual journey becomes infinitely significant as part of the larger awakening.

What about moral responsibility? This view may actually enhance moral responsibility. Understanding the interconnectedness of all existence makes you more aware of how your actions ripple through the web of relationships that constitute reality.

The practical path: This understanding suggests specific practices for aligning with the nature of what is:

1. **Cultivate present-moment awareness**—Since all is now, the present moment contains everything.
2. **Practice radical acceptance**—What is, is perfect as it is, including apparent imperfections.
3. **Navigate consciously**—Your choices matter precisely because they are part of what is.
4. **Recognize unity**—You are fundamentally connected to everything that exists.
5. **Embrace mystery**—The unknown is infinite and full of wonder.

What is? I am. You are. The universe is. The question is. The answer is. The infinite play of consciousness explores itself through every possible perspective by applying various combinations of sensory input and dimensionality, creating the magnificent complexity of existence in order to know itself completely.

You may be merely described as the nodal awareness and materialization of angular frequency interactions within an infinite expanse of vibrational energy. This may be what everything is. And it might be perfect, just as it is.

If, after processing this, existence is not perfect from your unique perspective, consider that you possess the universal power of consciousness to improve harmony within your sphere of influence. If you want something, work to understand it, know it can be yours, speak it into existence, take appropriate steps, and remember the old saying, "Be careful what you wish for." The universe provides because everything already is.

References

Barbour, J. (1999). *The End of Time: The Next Revolution in Our Understanding of the Universe.* Oxford University Press.

Chalmers, D. (1996). *The Conscious Mind: In Search of a Fundamental Theory.* Oxford University Press.

Einstein, A. (1905). On the electrodynamics of moving bodies. *Annalen der Physik*, 17(10), 891-921.

Goff, P. (2019). *Consciousness and Fundamental Reality.* Oxford University Press.

Lewis, D. (1986). *On the Plurality of Worlds.* Blackwell Publishers.

Maslow, A. H. (1943). A theory of human motivation. *Psychological Review*, 50(4), 370-396.

Padgett, J. (2023). Unveiling the cosmic code: Light, velocity, acceleration, and gravity in Quantum Information Holography (QIH). *Medium.* Retrieved from

https://medium.com/@jasonpadgett_76068/unveiling-the-cosmic-code-light-velocity-acceleration-and-gravity-in-quantum-information-db090f98ab5c

Tegmark, M. (2014). *Our Mathematical Universe: My Quest for the Ultimate Nature of Reality*. Knopf.

Thompson, E. (2007). *Mind in Life: Biology, Phenomenology, and the Sciences of Mind*. Harvard University Press.

Tononi, G. (2008). The integrated information theory of consciousness. *Biological Bulletin*, 215(3), 216-242.

U.S. Department of Energy. (2024). Visible light: Reading the rainbow for NNSA's missions. *Department of Energy*. Retrieved from

https://www.energy.gov/nnsa/articles/visible-light-reading-rainbow-nnsas-missions

Wheeler, J. A. (1989). Information, physics, quantum: The search for links. *Proceedings of the*

3rd International Symposium on Foundations of Quantum Mechanics, 354-368.

Wennerstierna, T. (2025). Quantum Information Holography and Consciousness: Toward a Unified Light-Based Model of Mind and Matter. Retrieved from https://tommywennerstierna.wordpress.com/2025/03/16/quantum-information-holography-and-consciousness-toward-a-unified-light-based-model-of-mind-and-matter/

Author's Note

This work presents an integrated philosophical framework that draws upon contemporary physics, consciousness studies, information theory, and direct phenomenological investigation. While grounded in established scientific principles, the metaphysical conclusions, represent plausible inferences that emerge from connecting diverse observations rather than proven facts.

The rich tapestry of perceptual limitations explored, from electromagnetic spectra to dimensional constraints, from temporal illusions to quantum mysteries, all points toward the same remarkable possibility: that reality may be far more unified, timeless, and consciousness-centered than our everyday experience suggests. The logical gaps between scientific observations and ultimate conclusions are acknowledged, and readers are encouraged to engage critically with the ideas presented while remaining open to the profound possibilities they suggest.

The examples and visualizations throughout are designed to help readers not just understand these concepts intellectually, but also to feel them viscerally, to see the limitations of their own perceptual bubble and glimpse the infinite mystery that lies beyond. For it is only by truly grasping the magnitude of what we don't and can't perceive that we open ourselves to the revolutionary possibility that everything, absolutely everything, might be here, now, perfect and eternal, waiting for consciousness to recognize its own infinite nature.

www.ingramcontent.com/pod-product-compliance
Lightning Source LLC
LaVergne TN
LVHW050938210926
848482LV00013B/664
9798891225770